INTO Wild Alaska

**BLACKBIRCH®
PRESS**

THOMSON
™
GALE

San Diego • Detroit • New York • San Francisco • Cleveland • New Haven, Conn. • Waterville, Maine • London • Munich

© 2004 by Blackbirch Press™. Blackbirch Press™ is an imprint of The Gale Group, Inc.,
a division of Thomson Learning, Inc.

Blackbirch Press™ and Thomson Learning™ are trademarks used herein under license.

For more information, contact
The Gale Group, Inc.
27500 Drake Rd.
Farmington Hills, MI 48331-3535
Or you can visit our Internet site at http://www.gale.com

Photo credits: cover, pages all © Discovery Communications, Inc. except for pages 6–7, 18–19,
21, 28–29, 32–33, 38, 40, 45 © Blackbirch Press Archives. Images on bottom banner ©
PhotoDisc, Corel Corporation, and Powerphoto; pages 26–27 © CORBIS

Discovery Communications, Discovery Communications logo, TLC (The Learning Channel), TLC
(The Learning Channel) logo, Animal Planet, and the Animal Planet logo are trademarks of
Discovery Communications Inc., used under license.

LIBRARY OF CONGRESS CATALOGING-IN-PUBLICATION DATA

Into wild Alaska / Elaine Pascoe, book editor.
    p. cm. — (The Jeff Corwin experience)
Summary: Television personality Jeff Corwin takes the reader on an expedition through
the harsh environment of Alaska, and introduces some of the diverse wildife found
there.
Includes bibliographical references (p.) and index.
  ISBN 1-4103-0059-5 (alk. paper) — ISBN 1-4103-0180-X (pbk. : alk. paper)
  1. Alaska—Description and travel—Juvenile literature. 2. Natural history—Alaska—
Juvenile literature. 3. Wilderness areas—Alaska—Juvenile literature. 4. Corwin, Jeff—
Journeys—Alaska—Juvenile literature. [1. Zoology—Alaska. 2. Alaska—Description and
travel. 3. Corwin, Jeff.] I. Pascoe, Elaine. II. Corwin, Jeff. III. Series.

F910.5.I57 2004
591.9798—dc21

2003009285

Printed in China
10 9 8 7 6 5 4 3 2 1

Ever since I was a kid, I dreamed about traveling around the world, visiting exotic places, and seeing all kinds of incredible animals. And now, guess what? That's exactly what I get to do!

Yes, I am incredibly lucky. But, you don't have to have your own television show on Animal Planet to go off and explore the natural world around you. I mean, I travel to Madagascar and the Amazon and all kinds of really cool places—but I don't need to go that far to see amazing wildlife up close. In fact, I can find thousands of incredible critters right here, in my own backyard—or in my neighbor's yard (he does get kind of upset when he finds me crawling around in the bushes, though). The point is, no matter where you are, there's fantastic stuff to see in nature. All you have to do is look.

I love snakes, for example. Now, I've come face to face with the world's most venomous vipers—some of the biggest, some of the strongest, and some of the rarest. But I've also found an amazing variety of snakes just traveling around my home state of Massachusetts. And I've taken trips to preserves, and state parks, and national parks—and in each place I've enjoyed unique and exciting plants and animals. So, if I can do it, you can do it, too (except for the hunting venomous snakes part!). So, plan a nature hike with some friends. Organize some projects with your science teacher at school. Ask mom and dad to put a state or a national park on the list of things to do on your next family vacation. Build a bird house. Whatever. But get out there.

As you read through these pages and look at the photos, you'll probably see how jazzed I get when I come face to face with beautiful animals. That's good. I want you to feel that excitement. And I want you to remember that—even if you don't have your own TV show—you can still experience the awesome beauty of nature almost anywhere you go—any day of the week. I only hope that I can help bring that awesome power and beauty a little closer to you. Enjoy!

Best Wishes!
Jeff

INTO
Wild Alaska

It's an incredible wilderness, teeming with wildlife—the only state with more square miles than people. The size of its unexplored terrain is mind-boggling. No wonder Alaska is nicknamed "The Last Frontier." We've got a great adventure ahead.

I'm Jeff Corwin.
Welcome to Alaska.

We're heading toward Denali.

Join me in a northward trek through Alaska as we seek as its largest resident, the giant and graceful Alaskan bull moose. The search will take us from the Kenai Peninsula on the southern coast, northward past Anchorage, around the towering peak of Denali, and into the heart of Alaska's largest national park.

Denali

Anchorage

Kenai Fjords

Alaska's rugged landscape is breathtaking.

These waters are teeming with life.

Our journey begins in the cold clear waters of the Kenai Fjord. These waters are filled with creatures like Dall's porpoises—the fastest swimmers of all porpoises, dolphins, and whales. They've been clocked at speeds of up to 35 miles per hour.

We've also found some sea otters, a mother and a pup. These animals do a lot of grooming because they have to incorporate lots of air into their dense fur. That air creates a nice invisible layer that traps warmth, and that helps them survive. Unlike other marine mammals, such as whales, they don't have a thick layer of blubber to stay warm.

Sea otters

Bye guys!

I could smell these animals before I could see them. They're Steller's sea lions. The number of these animals is dropping radically, and there's great puzzlement over the reason why. One theory has to do with the food they eat. Traditionally these sea lions like to eat fish that are high in fat, with very oily flesh—fish like salmon, for example. But a lot of those fish are scarce now, so the sea lions are having to rely on fish that are leaner, like pollack, with less fat content in their bodies. They're eating a marine version of junk food. When the young sea lions are weaned and head out to

I'm smellin' sea lions...

...And there they are.

sea, they don't have enough fat to survive. So the adults are reproducing in normal numbers, but the new generation is not surviving.

Here's a harem of sea lions.

The bull is the one in charge. An adult bull Steller's sea lion can reach 1,500 pounds in size, and more than 9 feet in length. The bigger he is, the more females he has; and the bigger the harem, the more successful he is. You can't mistake his sound. It's a low, rumbling growl, almost like a moan.

Alaska has 50 percent more coastline than the other 49 states combined. In all, 33,000 miles of land border water in this enormous state, making for some incredible geology.

Check out the chunks of ice.

I've traded in my Zodiac for a kayak so I can more easily maneuver through these large chunks of ice. Isn't this breathtaking? This huge ice sheet is about half a mile across, and at its highest point it's 600 feet tall. It goes back about 4 miles.

It's called the Aialik Glacier. *Aialik* means "a frightening or scary place," and I think that name comes from the continuous rumbling, growling, and cracking sounds this glacier makes as it dumps chunks of ice into the sea.

This glacier is constantly dumping ice...

You may think of ice as something that doesn't move, but this glacier is traveling about 4 to 5 feet every day.

While we're checking out the glacier, little heads are bobbing up, checking us out. About a hundred harbor seals are all around us in this area.

Checkin' me out.

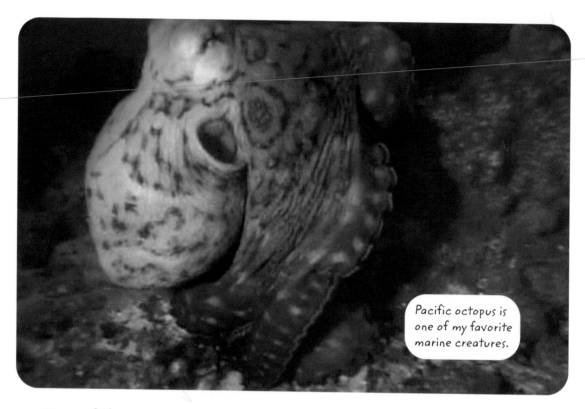

Pacific octopus is one of my favorite marine creatures.

One of the most mysterious creatures in these waters never worries about the ice. That's because it spends its entire life under the water. It's very rare to find one in the wild, but there is one living here at the Alaska Sea Life Center. It happens to be among my top ten favorite marine creatures—a giant Pacific octopus.

Even though the octopus belongs to the same group that clams and snails belong to, the mollusk group, it's extremely intelligent.

And man, he is strong. You see all his strength expressed in the suction discs that cover his underside and extend from each of his arms. He's curious—he wants to know who I am, and maybe find out if I'm food. Maybe he wants to show who's in control here. You can hear all those suction discs popping as his arms release. And he's sprayed me with his siphon. The siphon is a large, pulsating valve on the side of the animal's big domelike head. The octopus uses it for propulsion, to shoot forward like a jet.

By the way, they call this a giant octopus because it can reach a size, from tip of arm to tip of arm, of 20 feet or more.

Breakfast with puffins.

Cool birds

There's nothing like waking up and having a little breakfast. I like to start off with a few smelts myself, and so do these creatures. They're tufted puffins, beautiful birds, and they've made their home here at the Alaskan Sea Life Center.

See the tufts on the head?

The males and females look nearly identical. During the breeding season and the summer, when they're raising their hatchlings, their faces are pale and they have gorgeous plumes, or crests, which come out over their eyes. But when summer ends, their faces darken and they lose their crests. For months the birds will live out at sea. But they will return for the next breeding season and grow back those tufts.

A bird has to be designed in such way that flight comes easily. Now most birds have very light, hollow, porous bones. The puffin is a little different. A puffin does have hollow bones, but they're not nearly as porous as the bones of other birds. By having denser bones, the bird is able to stabilize itself better in the water.

Puffins are great swimmers.

Watch these guys eat—puffins just dive in and swim to the fish. Their vision underwater is excellent. They'll grab onto those fish and just gobble them down headfirst. Puffins are masters when it comes to swimming. They use the same motion of their wings underwater that they use in the air, rolling forward to create lift. I think they're at their most elegant when they're flying—not above the water, but in the water.

If all goes well, puffins can live about thirty years. And they're monogamous. A male and female stay together for their entire lives, and they put a lot of energy into raising their offspring.

Heading up one of Kenai's many streams offers us the rare opportunity to see one of nature's most extraordinary events. It's the running of salmon, an event that happens every year. And it shows us the ultimate price a species will pay to bring in the next generation. In the case of these pink salmon, that ultimate price is death.

Salmon heading upstream.

In the summer, the salmons' biological clock tells them to swim upstream to spawn. They live off their stored body fat as they undertake an arduous journey that can last

This guy's a male.

two to three months. It's easy to tell the sexes apart with these fish. The male's muzzle is almost pencil-like; compared to the female, he looks like a completely different species of fish.

When they reach their home streams, the salmon sweep a sort of bed, called a redd, in the sand. The female deposits her eggs there, and the male then puts his sperm on top of the eggs. These fish just totally exhaust themselves to make the

Many salmon don't survive the journey. Even those that do, die shortly afterward.

journey up this river to the areas where they're going to spawn. Many of these salmon are unable to complete the journey, and the ones that do will probably die about ten days later.

Where else but Alaska could you walk through a pristine river with water so pure you could drink it, teeming with salmon? Everywhere you step there's a squirming salmon. And after you've quenched your thirst, you can stop for breakfast. Everywhere there are blueberries, and I love blueberries.

Wild Alaskan blueberries...yum.

I think we're not alone...

Uh-oh—we've come across the remains of a salmon. That, my friends, is evidence that a large carnivore has moved through here, and my guess that carnivore was a bear. I have absolutely no desire to come face to face with that bear, so I'm going to head back to my Zodiac.

Yikes. There he is.

And there it is—a black bear, maybe the same one that chowed down on that salmon. See what this creature's doing? It's digging its way through this habitat, just looking for plants, berries, grubs, all sorts of things to eat. Often, when we think of bears, we think

These creatures are awesome— but I don't want to get too close.

of a creature that hunts. In fact, 75 percent of a black bear's diet is made up of plant matter. This bear probably weighs 300 or 400 pounds. It needs to consume up to 20 pounds of food today just to keep going. That doesn't include packing on weight for winter.

Black bears are perhaps the most common bear throughout the world, and certainly the most common bear in North America. They're also quite large. An adult male black bear can stand up to 6 feet in height and weigh easily 600 pounds. In fact, there are records of these bears pushing almost 800 pounds. Something else that's interesting about black bears is that they will occasionally hybridize, or crossbreed, with other species, such as grizzly bears.

Black bears are the most common North American bear.

The bear's reputation of being aggressive isn't completely deserved. This animal will defend itself and can reach speeds of up to 30 miles per hour and—although mothers are quick to defend their offspring—they are, for the most part, peaceful animals and would prefer to stay away from people.

One of the animals that bears like to hunt is the salmon. What's interesting is that the bear wants the most nutritious, the fattest, the richest part of the salmon. Now we think of salmon flesh as being rich, but to a bear the best part is the brains. Often you'll find pink salmon completely decapitated, the heads ripped open, and the brains gone.

Still heading north, we have to make our way through Big Game Alaska. This wildlife rescue center, not far from Anchorage, is home to a very interesting kind of deer.

A beautiful Sitka deer.

These animals are black-tailed deer, also called Sitka deer. And as you can see, they're not very large. In fact, the adult buck right here probably weighs some-where between 60 or 80 pounds. In addition to being one of the small-est types of deer in North America, these

Help! Deer attack!

animals have this interesting way to attack. They come in a pack, one from the front, one from behind, and—ouch! Big Game Alaska gives new meaning to the phrase "contact with wildlife."

Mating season means aggressive males.

There's another Alaskan native just up the road, and this one is making a comeback after being hunted to near extinction in the late 1800s. The facility houses a herd of musk ox. This is the rutting season, when males become extremely aggressive and volatile. The females are much more docile, but still dangerous to people, so I'll have to watch my step.

Hey, guys.

The primary defense of these animals when they encounter predators such as wolves is to move the calves

into the center of the herd. The adults stand around the outside, locked together, and go face to face with the predators. They'll keep a pack of wolves at bay with their great horns, while the younger ones are safe in the center.

Under the outer hairs lies the dense, fine fur.

One of their best defenses from the cold is buried underneath their coarse outer hairs. It's an insulating layer of fine, dense fur called qiviut. In fact, these animals are part of a unique project—a sustainable harvest of qiviut. When this dense fur sheds out, it's collected, spun into yarn, and woven into fabric. Qiviut is so dense, so luscious, so warm, that it's eight times denser than sheep's wool and much finer than cashmere.

These horns are like daggers.

I'll keep my distance from this bull.

The musk ox has very dramatic horns, and the best way to check out those horns is by looking at a bull. Look at this guy—he's huge, six years old and about 1,000 pounds. He's a very dominant bull and getting into his rut stage. This is the point in the year when he's the most unpredictable, the most aggressive, and the most ornery. You really don't want to mess with him.

The horns on top are like solid rock.

Horns actually grow like hair.

I'm just blown away by this animal's horns. Check out the boss, the part of the horns that protects his head. It is like a battering ram, solid as stone. Those horns look like they're made of bone, rock solid and tough. But the material is more related to hair—the horns grow like hair and have the same chemical structure as hair.

If this bull were in the wild, his horns would grow longer. They would actually curl around and be very sharp at the end, making him even more dangerous. Imagine what it's like when these bulls are battling—two 1,000-pound animals smashing together, with a sound like thunder. It could do a lot of damage to this animal's body. But he has a huge muscular neck that can absorb all the shock when he smashes against a competitor.

This is Denali. *Denali* means "the high one" in the Athabascan native tongue, and it's the original name of Mount McKinley— the tallest mountain in North America, at 20,000 feet. Seventy-nine people have died climbing this behemoth, mostly due to the treacher-ous weather.

Denali is the tallest mountain in North America.

Temperatures near the peak can drop to 95 degrees below zero, with winds of up to 150 miles per hour.

Lookin' for pika...

Denali is also the name of the national park that surrounds the mountain. It is an enormous park, larger than my home state of Massachusetts. This rocky outcrop is great for finding a very neat lagomorph, a relative of the rabbit, called a pika. Just look in these little crannies to find it.

These little pikas are food for many predators, everything from birds of prey, to foxes, to wolverines, to you name it. What's neat about this little guy is that he takes everything he forages and piles it into a cache. His cache is made up of nuts, seeds, grasses, and dried plants, and he'll live on that through winter. He does not hibernate at all. He weighs less than a can of soda and

Pikas are in the rabbit family.

could potentially, if he's a really good collector, put together a cache of anywhere from 12 to 50 pounds of food. That's pretty impressive.

This guy's like a little electron. He just darts around. It's hard to keep track of him.

We have wandered into the territory of a pack of wolves. I've been told that there are six members in this pack, and they wander through the region of Denali in search of prey. They could be all around us. It's very unlikely we'll see them, but we know they're here.

We're in wolf territory.

Right there. Wolf tracks.

Here's a surprise—a red fox.

Here's an animal I didn't expect to see—a beautiful red fox. He just woke up. He's got a den back behind these rocks. He's what we'd call a subadult, probably around four months in age and just reaching that point where he's getting out and exploring. This guy doesn't

Look at the color of that fur. Gorgeous animal.

seem to have much fear of us. He's smelling my backpack. Never in a million years did I think I could creep up to be 8 feet away from a wild fox. It doesn't get better than this.

How's this for a weather change? This morning I woke up and it was the end of summer, and now afternoon has come and winter has arrived. You may not see it, but you sure can feel it right to your bones. It's about 30 degrees. In Alaska, summer lasts as long as a bug in a bug light.

Instant winter.

Look at all that snow.

We're going up this steep slope of snow-covered moss, because at the top I've spotted a very interesting type of animal. I don't know if we'll catch up to it, but let's give it a try.

Spotted these Dall's sheep up on top.

Don't want to scare them away.

About 40 feet ahead of us is a herd of Dall's sheep. I see two rams and four or five ewes; they have the smaller horns. Look at that. You can actually see a young lamb hanging out with one of the rams. There's one female who's acting as a sentry, checking us out.

We've climbed all this way and we're so close; yet I have a feeling that if we get any closer, they're going to take off. And these animals are just masters when it comes to climbing. They've got cloven hooves, which dig into the rocks, and they can just shimmy

These guys are
expert climbers.

their way up to
the peaks. And
in winter they
can go clear up
thousands and
thousands of feet
in altitude and
survive where any
meager man, such
as myself, would
perish instantly.

Whoa! Look what I found!

I'd like to follow these guys some more, but I'm frozen. Time for a hot toddy.

Wow. See that? It's a grizzly bear. I've got to find a way to get closer to it, but stay at a safe distance.

I can't believe it—a grizzly.

Grizzlies like to live alone.

Grizzlies are, for the most part, solitary animals, and a healthy bear often lives for up to twenty five to thirty years in the wild. Just look at him. He's having a ball in this early snowfall. We are lucky to spot this guy. Winter's coming and soon he will be sound asleep until spring.

Our northward trek continues, and the moose can't be far. I've found a nice fresh pile of moose scat, still warm and aromatic and steaming. I just love these little gifts from nature.

I see something in the distance...

And look at that—there's your quintessential Alaskan postcard, a beautiful green swamp with a lagoon. And feasting away is a very large moose cow, probably about eight or nine years in age. She's got two calves, probably about four or five months in age.

We're seeing a very successful cow. Perhaps she's successful because of her genetics. Maybe she's successful because of her parenting skills, but the expression of her success is that she has two offspring and both have survived to this age. That's very rare.

And there's an excellent chance that in about six months they'll still be alive, and they'll be able to take the challenges of this environment on their own.

I love these moose.

During the birthing season, moose cows drop their calves all at once, flooding the environment with hundreds and hundreds of babies. If they were to drop their calves one by one over an extended period of time, all those babies would be snatched up by predators. But this way, predators are overwhelmed and can only take a fraction of the calf population.

All right, we found our cow moose. Now, let's see if there are some bulls hanging around here as well.

Just over there you can see, hanging over the brush, a rack— a gorgeous rack—and it's connected to the animal we've come to find. Up here in Alaska, the moose give new meaning to the word

Look at that rack...

gigantic. Think about a moose, and you may think of something clumsy or silly—like Bullwinkle. That moose is nothing like this moose.

You'd think having all that great body mass would slow you down, but that's not the case for a moose. They're extremely fast when they need to be. And this guy, if he felt threatened, could easily charge me at 30 miles an hour. That's a lot faster than I can run. This is the time when bulls are pretty much at their most aggressive state, so we should move in very carefully.

We've got some exciting stuff happening here. Look at this! We have a cow moose, and the cow has brought in two suitors, two bulls.

The lesser bull doesn't want anything to do with the bigger bull. But the bigger guy is ticked off that the smaller guy is in his territory, so he's on his tail, pushing him out. The bigger bull came down and said, "You're in the territory of my cow." And that lesser bull muttered a submissive whine that said, "Okay, hey, no hard feelings."

Look at this scene. This is what Alaska's all about.

Males will fight during mating season.

When bulls are more evenly matched, they'll fight for their territory and for access to moose cows. Rarely witnessed by people, it's often a struggle of titanic proportions. The bulls will spar until one gives up and the other claims victory and possession of the cow.

This has been fantastic. Alaska, thank you for giving us this very exciting encounter. One last picture and I'm off. I'll see you on our next adventure.

A parting shot...

Okay, so I'm a tourist, too...

Glossary

arduous difficult

blubber a thick layer of fat in marine mammals

boss the part of an animal's horns that protects the head

buck a male deer

carnivore an animal that eats meat

cashmere fine wool from a particular kind of goat

cloven hooves animal's feet that are divided in two parts, such as on a goat

docile calm and easily led

ewe a female sheep

extinction when no more members of a species are alive

forage to wander and search for food on the ground

hybridize to crossbreed with another species

lagomorph a type of mammal that includes rabbits and pikas

mollusk a type of animal that includes snails, clams, and octopuses

monogamous animals that mate for life

predators animals that kill and eat other animals

pristine clean and unspoiled

qiviut the layer of fine, dense fur on a musk ox

quintessential the most typical example of something

rutting season an animal's mating season

scat animal droppings

siphon a valve on the head of an octopus

smelt a type of small fish

spawn to lay eggs for reproduction

volatile subject to sudden change, explosive

Index